In-Space Robotic Repair and Servicing of Spacecraft

Patrick H. Stakem

2nd edition

Number 14 in the Space series

(c) May, 2019

Table of Contents

Introduction..5
   The author..6
A note on Units..7
Why service in orbit?..8
      Servicing at LEO 9
      Servicing at GEO, and the Geo Graveyard...........................11
      Robotics on the ISS...14
      Canadarm..16
      Candarm2..16
      Servicing in Polar Orbit...19
      Servicing at L2 19
      Lunar Surface Servicing..21
      Mars in-orbit and on-surface servicing...............................22
      Asteroid...23
   Design for Servicing...23
   Enabling technologies...24
     Tactile sensing..25
     Telerobotics versus robotics...26

NASA Experience 30

   Rendezvous, Docking, and Berthing............................30

   GSFC's Multi-mission Modular Spacecraft....................33

   The FSS...34

   The SMM Repair Mission..34

   The Hubble Repairs..36

   NASA-GSFC SSCO, Satellite Servicing Capabilities Office...39

      RESTORE Project..40

      The OMV...43

      Robotic Refueling Mission....................................43

      NASA's DART...46

The Military Experience 46

   DARPA Orbital Express...47

      NextSat..48

      Astro..49

The Commercial Spacecraft Experience...............................49

   Intelsat 50

   Orbital-ATK..52

International efforts..53

ESA..54

German Efforts...54

Canadian Space Robotics.................................55

Other Approaches...55

Swarms...55

Wrap-up...56

Bibliography..58

Resources..71

Glossary...75

If you enjoyed this book, you might also be interested in some of these..81

Introduction

This book covers the topic of On-orbit repair and servicing of spacecraft. Putting a communications satellite in synchronous orbit will set you back 100's of millions of dollars. Once on orbit, you hope it survived the launch environment, and operates correctly. You further hope it works at least for its design lifetime, and as long as possible. This approach, based on good engineering design practices, lessons learned, and hope, it the equivalent of buying a new Tesla with non-rechargeable batteries, and driving it until it stops. Then buying a new one.

We will discuss the history and the technology of on-orbit servicing, and the projects currently being conducted. We'll take a look at ambitious planned projects, and the enabling technologies that will make them a success. We'll speculate what this means to missions to other planets in our solar system, and the challenges to manned expeditions to follow the robotic ones.

When Columbus sailed to the new world, he had to bring everything he needed with him, because he was literally sailing into the unknown. If he couldn't find sources of food and water, he would have to turn back. The sailors knew how to fish, and how to trap rainwater in sails, so they were ok for a while. If something happened to the ships, they had to rely on the skill of the ship's carpenter to fix it. They could not be sure that could find wood they could use, so they would have spare spars, rigging, pulley blocks, and other necessities. Once it had been established that there was land there, it made the journey a little less perilous. The fact remained that repairs had to be made during the journey, and, at sea, they had no supplies and tools beyond what they carried with them.

This book was recently updated, as a Northrop Grumman spacecraft successfully docked with a depleted communications satellite in geosynchronous orbit, allowing it to be placed back into service.

The author

The author spent his career in support of numerous NASA spaceflight mission, on this planet and others. He has taught for the graduate Computer Science Department of Loyola University in Maryland, Capitol Technology University, and the Johns Hopkins University, Whiting School of Engineering, Engineering for Professionals Program.

He spent 42 years as a NASA contractor, at all of the NASA sites. He began his aerospace career at Fairchild Industries, and overlapped with Dr. Wernher von Braun, so, technically, he was a member of the von Braun rocket team. He has received the NASA Shuttle Program Managers Commendation Award, two NASA Group Achievement Awards, Certificate of Appreciation from the NASA Earth Science Technology Office, and the

NASA Apollo-Soyuz Test Program Award. He did extensive work on the Flight Telerobotic Servicer Mission. He served on AIAA's Committee on Standards for Automation and Robotics for TransOrbital, Lunar and Mars Base. He worked on the Solar Maximum Repair Mission, and briefly on the Hubble Space Telescope repair missions, as well as other Shuttle missions.

Mr. Stakem is on Facebook and Linkedin.

A note on Units

I am fairly conversant in both English and Metric units (what is the metric equivalent of furlongs per fortnight?). Metric (SI) is mandated for NASA usage now, for interchangeability with our partner space faring nations. When a lot of the legacy flights discussed here were flown, English units were the norm. I have tried to keep the units comparable to the mission at the time. Conversions are easy enough, but units conversion is a source of error. It's not what you know about units and measurement, its how you think. And, I still think English units (even the English use Metric now), and convert in my head or on my phone.

For scientific/engineering work, the Metric system is well thought out. For artisans, the English system served well, as most units were divided by 2. Which is easy. Fold the cloth. Hopefully, when we are all taught Metric first, some one will still remember the conversions. You just need a good slide rule....

Why service in orbit?

Serviceability provides a flexibility at the cost of complexity. This is a step beyond reconfiguration from the ground to bypass errors and use redundant equipment, or reprogram around the problem.

Refueling is a mission-extending approach - spacecraft have propellant for maneuvering, orbit maintenance, and attitude control. Another limiting factor for spacecraft life might be onboard helium, which is used to super-cool certain detectors in the instrument packages. Liquid fuel, from a dead satellite or leakage, freezes in space and provides yet more debris problems. There are many otherwise functioning satellites that have simply run out of fuel. The hardware has exceeded the lifetime projections, and a very expensive asset is now useless, and actually a hazard.

Sometimes, even multiple redundant systems fail. In that case, it is essential to have a serviceable architecture to allow repair.

The capability to service in orbit brings with it the capability to assembly large payloads in orbit, for even more advanced missions to other planets. The ISS was assembled in orbit, and the planned Deep Space Gateway will be.

In another scenario, satellites are sometimes put into the wrong orbit, due to problems with the booster. Their orbit is not compatible with their mission, and they are essentially useless.

Commercial and government missions are different. The government is self-insured, but commercial payloads are usually insured. This means the underwriters need to be up to speed on what can fail, and when. With payload and launch costs in the hundreds of millions of dollars, this is a big deal. The commercial satellite operators are not in the business of launching payloads to gain new knowledge about the universe. They want to provide services for which people and other companies will pay (think, Dish Network). It the satellite fails to achieve orbit, that is an expensive asset lost. If it runs out of fuel earlier than anticipated, that becomes a non-performing asset.

Servicing at LEO

Spacecraft servicing at low Earth orbit has been done by Astronauts in EVA operations from the Space Shuttle. We will discuss these cases later on. The Shuttle Fleet has been decommissioned, but the lessons-learned are invaluable.

When low Earth orbiting satellites fail, they can be reentered into the atmosphere. In fact, this capability is now required for all new missions. This frees up space in orbit, and also removes what is now a big chunk of debris, that could endanger other spacecraft or the Space Station.

Space debris, from failed satellites to nuts and bolts, pose a problem to other satellites. A bolt traveling at 18,500 mph is a big issue. There have been 5 known collisions of satellites in orbit so far. Periodically, the ISS has to do a debris-mitigation maneuver. All of the junk, down to the size of bolts, is tracked by the U.S. Air Force. They know of 18,000 objects in orbit, of which 1,400 are operational satellites. A good job for a robotic servicer in LEO would be to collect the trash, put it in a canister, and kick it off to re-enter and burn in the atmosphere. There are estimated to be 170 million chunks, smaller than a centimeter, any of which can ruin your mission.
Known space debris includes Astronaut Ed White's outer glove, lost on his space walk; Michael Collins's camera from Gemini-10; a wrench, pair of pliers, a tooth brush; and a complete tool bag from an STS-26 EVA.

For mitigation, the ISS uses Whipple shielding, named for its inventor. This uses a thin outer bumper, spaced away from the hull. The idea is, the bumper breaks up the debris so that penetration doesn't happen. It works most of the time. Think of it as the ISS's bullet proof vest.

And, the Vanguard-1, launched in 1958, is still in orbit, and so is it's upper stage. I think it would be a great crowd-funded project to bring it back, and put it in the

Smithsonian. We put it in space, we should be able to get it back.

Servicing at GEO, and the Geo Graveyard

Although there are a lot of scientific and imaging satellites in Low Earth Orbit, up to a few hundred miles, but the communication satellites' preferred spot is in synchronous orbit. You will also find the Tracking and Data Relay Satellites (TDRS) and a lot of the weather satellites (GOES) there. This is some 22,300 miles high, where the rotation rate matches that of the Earth, and the satellite maintains the same east-west position. Actually, this is only true over the equator, and is hard to achieve and maintain. Synchronous spacecraft usually wander north and south of the Equator in a lazy figure-8 pattern, still within the beam of the Earth-based antennas. There are about 400 satellites currently occupying geosync orbits. When they fail, or run out of fuel, they are turned off, put into the graveyard orbit if fuel and functionality permits, and left to rot. Thousands of years from now, they will still be there, in the same condition. There are about 100 spacecraft there now, and we're adding about 20 per year. As more spacecraft join them, the probability of collisions increase.

The graveyard orbit near geosynchronous is some 100-300 km higher to ensure the dead satellites do not interfere with the operational ones. Roughly, 20 spacecraft at GEO expire each year, after a 20 year operational life. Unfortunately, the graveyard orbit is not stable in the long run. The moon is the problem. The moon's effect causes the orbits to be perturbed slowly over time. In geosync, active spacecraft adjust their orbit periodically with thruster firings. This results in the requirement for refueling. In the graveyard, the dead spacecraft can't.

After a collision, the problem gets worse. Now, instead of two spacecraft, we have hundreds of spacecraft pieces. The energy of the collision disturbs the orbits, and each piece is going off on its own. Space is a big place, but the probability of more collisions becomes greater. As does the possibility of pieces reaching and damaging active spacecraft. Not a good long term solution.

There are two solutions. One is to service the spacecraft at geosync. This means they will have an extended lifetime. For the spacecraft in the graveyard orbit, they could be put into a solar intercept orbit, as the pieces of the Saturn-Apollo rockets were. In the short term, we should focus on keeping what's up there in good running order, with lots of propellant.

This has two aspects. The spacecraft launched to geosync should be designed to be serviced. This makes the operation much easier. We'll discuss what this entails later. Secondly, we need a service vehicle to get to geosync, autonomously service and repair, and return, probably to the space station.

NASA and the DoD are both working on an approaches to servicing at geosync. GSFC's Satellite Servicing Capabilities Office (SSCO) has a Rotational Robotic Servicing Mission in definition. Other commercial companies are working on the problem as well, as it will become a lucrative business, "this concept would bring a gas pump, mechanic, and tow truck to satellites in space."

An orbital propellant depot is a proposed resource for satellite life extension. Think of it as a neighborhood gas station. Those company's with geosync resources have shown interest, but the project hasn't gotten any traction. A lunar or Mars mission could launch with minimal propellant to Leo, then top up the tanks before setting off. The best propellant, liquid hydrogen and oxygen, are cryogenic, stored as a liquid. These need to be stored in super-insulated tanks, to minimize boil-off. Other fuels are less demanding, but also produce less energy. Extending the idea, propellant depots could be put at the Earth-Moon L2 point, or in lunar orbit. We may find and exploit water ice on the moon.

Liquids of any type are problematic payloads. As some of the liquid is consumed, there is free space in the tankage. The liquids slosh back and forth, presenting problems for the launch vehicle control system. This is mitigated by internal baffles, and possible gas bladders.

In any case, the ullage, the last remaining amount of fuel, is usually hard to access.

Robotics on the ISS

This section discusses the use of robotic and telerobotic systems on the International Space Station. On-orbit construction of the ISS began in 1998, and was completed with a last Shuttle mission in 2011. It is the largest artificial satellite in Earth orbit, and can be seen from the ground with the naked eye. The ISS is a synthesis of several space station modules from the U. S., the Soviets/Russians, the Europeans, and the Japanese. It serves as a laboratory, observatory, and factory in Earth orbit, and is continuously crewed. The station currently has multiple robot arms as well as a mobile crane.

The Mobile Servicing System (MSS) on the International Space Station was produced by MDA Space Missions as a contribution from the Canadian Space Agency to the International mission. It arrived at the Station in 2001. It can move equipment and EVA astronauts around the outside of the station. The system consists of a mobile base that moves on rails outside of the station, on the Integrated Truss Structure. The base is called the Mobile Transporter Cart. The cart transports the Canadarm assembly, which may be itself connected to the Dextre robot, or an EVA astronaut.

The Mobile Servicing System consists of Candarm2, the Mobile Remote Servicer Base, and the Special Purpose Dexterous Manipulator. With the Shuttle-derived latching end effector, the system can latch onto and move large assemblies out the station.

The Special Purpose Dexterous Manipulator (SPDM), called Dextre is manufactured by the Canadian firm MacDonald Dettwiler. It is used in conjunction with the Canadarm assembly, and is sometimes referred to as the Canada Hand. It includes 2 short arms and various end-effectors for servicing operations.

It arrived at the Station in May of 2008, on shuttle Mission STS-123. It was assembled in orbit by EVA astronauts from the Station. It can be operated by the onboard crew in telerobotic fashion, or from the ground.

Dextre's arms are 11 feet long, fitted to a body assembly that can bend at the waist. The Canada arm assembly can grasp the Dextre body by means of a standard grapple fixture attached to its back. The ARM can be moved to various worksites around the Space Station. Each of Dextre's arms have an ORU/Tool Change-out Mechanism, that allows various tools to be attached. These include grasping jaws, a socket drive, a camera, light assemblies, a connector to provide power,r and a data connection. The robot body includes lights and dual color cameras, as well as a platform to hold ORU's and tools. The actual hand that attached to the Dextre arm is called SARAH, the Self Adaptive Robotic Auxiliary Hand. Dextre gets electrical power from the Canadarm2.

Canadarm

Formally called the Remote Manipulator System (RMS), each Shuttle carried one, and the Space Station has a updated version.

Candarm2

The derivative of the Shuttle's Arm, the Remote Manipulator System (RMS) is the Canadarm2 for the Space Station. It was launched in April of 2001. It is 58 feet long, and has seven degrees of freedom. It is capable of manipulating payloads of up to 256,000 lbs on orbit (where inertia, not weight, is the issue). The Arm interfaces with Power Data Grapple Fixtures (PDGF) located on the station. It can move from one PDGF to another , as each end of the arm assembly can be connected. It can also be moved on the Space Stations's railroad, the Mobile Base System, which rides on rails. Inside the Station, crew members at one of three Robotic Work Stations (RWS) located throughout the station, can view and operate the arm. It is operated by dual hand controllers, one for translation, and one for rotation.

The Japanese Experiment Module (JEM, named Kibo) is the largest on the station. It came up on three Shuttle missions. There are six major elements, including the pressurized lab, the exposed facility, and a robot arm.

The Robonaut is a circa-1997 NASA Johnson Space Center Dexterous Robotics Laboratory Project to define an Astronaut-equivalent humanoid telerobot for use inside or outside the Space Station. The focus for Robonaut is in dexterity, and safety in working with Astronauts. Being human-form, it can use tools developed for astronauts.

The initial Robonaut design, circa 1996, was to be used as an end-effector on the Station's robotic arm, so it could accomplish EVA tasks. Two versions were built, but none were flown.

The new version, Robonaut–2, was launched to the International Space Station onboard Space Shuttle flight STS-133 in February 2011. His legs were sent up on a later resupply flight, and were attached in 2014. Try that with an Astronaut. The Robonaut became a permanent resident on the station.

As an Astronaut-equivalent, the robot will have roughly the same size, strength, and dexterity as an Astronaut. It will use the same tools, handholds, hatches, and such. It has 5-fingered hands with 12 degrees of freedom, which are gloved-hand equivalent.

One or more Robonauts can perform co-operative tasks with astronauts. This aspect has been tested extensively on Earth at JSC. There are over 350 sensors in the Robonaut, and 38 PowerPC computers. The Robonaut is designed to be connected to a station laptop. It currently must be plugged into a station power outlet, but a new battery pack for the unit is being supplied by Boston-Power, with 2.5 kW-hours of capacity.

The unit weighs 330 pounds on Earth, and has a mainly aluminum structure. There are a total of 42 degrees of freedom in the unit, including 3 in the neck, 7 in the upper arm and wrist, and 12 in the hands. It also has waist rotation. Joints are controlled by servo motors. The fingers have tactile sensing, and integrated load cells in the finger joints. The finger gripping surfaces are a high friction material

Robonaut's may find additional off-planet work as explorers - keep an eye on this technology. On Earth, a large number of technologies from Robonaut are available to use under license. http://Robonaut.jsc.nasa.gov

The follow-on is Robonaut2, a joint NASA, General Motors, Oceaneering effort. It incorporates a lot of lessons-learned from the original Robonaut. The spins-offs in the areas of vision, image recognition, sensors, manipulator design, and control algorithms, results in more than 50 patents.

The major subsystems include the hands (manipulators), the arms. Sensing and perception, and interface and control. The unit can be teleoperated, or function autonomously. Then hands are designed to be able to use human tools.

Servicing in Polar Orbit

The plane of a polar orbit is highly inclined with respect to the plane of the equator. We speak of a equatorial orbit as having an inclination of zero. A true polar orbit would be inclined 90 degrees, and pass over both poles. In practice, polar orbiters have a small angle of inclination. If you get the math right, the spacecraft will pass over every spot on the Earth's surface periodically. This is useful for weather satellites. It takes a lot of energy to launch to Polar orbit, and a servicing satellite may not be needed yet, due to the low density of spacecraft currently in that path.

Servicing at L2

I am going to skip two semester's of graduate school math here, and just give you the results. The interested reader can do the math himself/herself, starting with f=ma. If you just consider the Earth and the moon, there is a point between them where the gravitation force is equal, and, theoretically, if you put something there, it will stay. In practice, it's not that easy. Interactions from other planets such as the big ones- Jupiter and Saturn, cause perturbations. In fact, you won't be able to calculate exactly what is going on. This is the 3-body Problem, and it has no know closed-form solutions. Luckily we can approximate a solution, that's "good enough." Well, it turns out that there are actually 5 positions in the restricted 3-body problem that are null points in the combined gravity field. Put something there, and it stays. Almost. But, it's well enough understood that we can put a satellite at the null position, and keep it there with some thruster activity now and then. Why are these null points, called the Lagrange points, important? We put sentry satellites at the one between the Earth and the sun, and monitor for solar storms. We can put a space telescope at the Sun-Earth L2 point, and make a decent observatory. This is the James Webb Space Telescope (JWST) Project, due to be launched in 2021. The Earth-Moon L2 point, behind the Moon, would be ideal for a communications satellite and a lunar observatory on the back side of the moon. Just to add one more note of conceptual complexity, the small object does not need to be placed exactly at the Lagrange point. It can orbit it. OK, admittedly a little strange. Orbiting a point in space with no primary. Accept that, or do the math, but I'm moving on.

The very expensive James Webb Space Telescope (JWST) will be placed at the Sun-Earth L2, almost

1,000,000 miles away. If it suffers a problem like the Hubble Space Telescope did in Earth orbit, there is currently no feasible way to service it, refuel it, or bring it back. The current costs of the program are estimated to be in excess of $8 billion. We actually won't know if its working until it gets there, and we hope our best engineering practices are effective. Could we service JWST at L2? In theory, yes. It is, as the Hubble was, not designed for servicing. It would be the same baseline as a servicing mission at synchronous Earth orbit, but a lot farther away. But, if that servicing could be accomplished for significantly less than the replacement cost, it might make sense, and enable the continued flow of science data.

Lunar Surface Servicing

Compared to in-orbit servicing, servicing on the surface of the moon would be easier. Gravity is our friend. We can actually see things on the (front) lunar surface from Earth. We know how to build Planetary Rovers from our Lunar and Mars experience. It might even be possible to retrieve items from the lunar surface and return them to Earth. We did that with Apollo. I watched that on TV.

Telerobotics suffers from the delay problem, when the device does not respond immediately to human input. Operating telerobotics on the surface on the Moon is frustrating, due to the near 1 second time delay involved. However, it is very possible to operate lunar surface robots from a position closer, such as the Deep Space Gateway. This project is for a crewed space station between the Earth and Moon. It could be used to control exploration and mining robots on the lunar surface. Of interest is lunar ice, which could provide fuel and oxidizer for deep space missions. The water ice is broken down into hydrogen and oxygen by solar powered electrolysis. The Saturn-V launch vehicle used liquid hydrogen and oxygen.

Mars in-orbit and on-surface servicing

Not to give too much credit to the Martian Planetary Defense Force, but around half of the attempted Mars missions fail. Some miss the planet, some get stuck in Earth orbit, some crash onto the surface. Once they land, we have had spectacular success with the Rovers. Just as important is the infrastructure. On Earth, we can assume GPS positioning, overhead views, a decent weather model, and communication satellites. We are beginning to build that infrastructure around Mars. Communications relays mean the lander does not have to have a radio system capable of reaching Earth. The "eyes in the sky" can see approaching sand storms, that would cover the solar panels with dust. Mars doesn't have a usable magnetic field to navigate by.

We could image a garage facility for Martian Rovers. We might need to include a "tow truck" if the rover's mobility is impaired. So, what does that Martian "AAA" buy us? The very expensive rover with its very expensive suite of science instruments could be waylaid by any of a number of problems. Most of these would stem from equipment damage or failure. Again, the rovers are not designed to repair (this is a matter of access and connectivity, as you might notice under the hood of your car). A standardized rover platform designed with ease of maintenance in mind could serve as a mobile base for instruments, as well as the base for the repair/recover vehicle.

It can be argued that several Mars missions have been repaired in situ, but this was done operationally, or via software update. If a wheel freezes up or falls off, the mission is probably over.

Asteroid

These idea can be extended to explorers of the outer planets. Given the very high cost of getting there in the first place with a science package, we can seen the advantages of a repair/refurbish capability. The problem with fixing things remotely is the limited number of things that can be fixed, mostly switching redundant units, power cycling, and updating software. But, one step at a time. We need to solve the problem for the thousands of Earth orbiters first.

Design for Servicing

It is much easier to service a device that was designed with servicing in mind. Up to a point, spacecraft were designed to service the launch environment, and hope for the best. The Shuttle provided a platform to retrieve and repair spacecraft, opening up a new realm of possibilities. Design for servicing involves modularization, standardization, and access.

Standard power and data connectors and fluid connections, and standardized and captive fasteners are part of the solution

A Grapple fixture is a standardized way for spacecraft to be handled on-orbit by astronauts, and the robot arms on the Space Shuttle and the Space Station. The Standard for the spacecraft side is a passive fixture, attached to the spacecraft, and including an optical target, and a central pin. The end effector of the arm assembly maneuvers into position, closes on the pin, and a cable is wound tight to lock the connection together. The standard grapple fixture was developed at Spar Aerospace, a Canadian Company. An advanced type is the Power Data Grapple Fixture, which includes data and power in a connector. Unmanned resupply missions to the Space Station, such as the Dragon Capsule, have a grapple fixture. This is used to capture the capsule with the Station's arm.

Enabling technologies

Robots are handicapped in terms of mobility and manipulation, sensory input, cognitive processing, learning and the application of experience. However, they have better computational capability, better communications capability, fewer environmental constraints, and, perhaps, fewer ethical issues. (Leaving aside the issue of military armed robots).

What are the problem areas in the applications of servicing robots? Accuracy, which has both a mechanical and a sensor component; dynamic performance, a speed/dexterity trade-off, addressed by more robust control algorithms, sensor systems in terms of integration and procession; interactive control, starting with modeling, standardization and modularization. None of these are insurmountable problems, and the advance of technology addresses all.

Robotic systems need better world models. They need to integrate and fuse sensor data into a better view of the world around them. They need more reasonableness assumptions, and a-priori knowledge of the physical world. They need flexibility of response. Learning from experience would be a major asset. What they need, then, is better software. The ideal component, it doesn't weigh anything.

Tactile sensing

The sensor model for the human skin is one of high but varying sensitivity, highest near the fingertips. It has a fast response an continuous output. It is flexible and durable, yet self-repairing. It is a smart sensor, containing a level of processing. Human skin can detect pressure, giving contour information; slippage, the pressure across a series of points; and temperature. This latter property allows for a level of materials identification by their thermal properties.

Telerobotics versus robotics

The word robot is from the Czech *robota*, which means servant or laborer. It was coined by novelist Karel Capek in a 1917 short story. His 1920's play R.U.R, Rossum's Universal Robots, brought the term to the public eye. "Robot" was first applied to describe a manipulator systems for manufacturing and the science fiction creations. A robot is a tool that is flexible and programmable.

One definition often used is that a robot is a "programmable multi-functional manipulator designed to move materials, part, tools, or specialized devices through variably programmed motions in the performance of a variety of tasks." It is completely autonomous once trained so that it can operate without further human intervention. It incorporates feedback in its operation. Telerobots increase the domain where useful work and observations can be done. They are human proxies in a hazardous environment, such as space.

"Tele-" is from the Greek, and means distant. A telerobot has a person in the control loop. A variation of this is a much smarter telerobot, with Directed Autonomy. There is a spectrum of sense-control tasks, and as more of these are assigned to the telerobot system, it becomes more autonomous. Well-structured tasks can be accomplished autonomously, after training.

Telerobotics are robotic devices which incorporate mobility, manipulative and sensing capability, and are controlled remotely by a human. The issue is, the level of control. The human operator might individually operate each joint or mechanism, or he/she might just say, "go explore." Telerobotics provide feedback to the operator, visually and hopefully by force-feedback. This leads to the mini-master model of hand control of manipulators in telerobots, developed in the nuclear industry. A telerobotic system, with a person in the loop, uses the best of both subsystems: the decision making and visual acuity of the human with the strength and dexterity, not to mention the ability to operate in hazardous areas capability of the robot system.

Telerobotics have been used in a variety of applications for more than 50 years. The nuclear industry has the longest history of operational experience. Telerobotic systems operate in toxic chemical and biological environments and in research. Bomb disposal is an obvious area, with units deployed with local police forces now being common. Remotely piloted vehicles are another area of telerobotics, and combat robots are increasing being deployed. Telerobots serve as prison guards and warehouse security. Some trains are telerobotically operated within the confines of industrial plants.

But teleoperation is tiring for the human operator. It generally requires the operator's full time attention, and may involve confusing communication delays. With supervisory level control, the operator is relieved from the tedious lower-level details, and can concentrate on goals.

However, depending on the capability of the robotic system, the worksite or task must be well structured, or the robot component must have extraordinary complex sensor processing and decision-making capabilities. But, the rapid advance of technology provides paths to making robotic systems smarter and more capable. This evolutionary approach allows a human operator to teach the robot a series of increasing complex tasks, which become subroutines that the robot can execute autonomously. As task analysis and planning become more automated, the role of the human is relegated to the upper levels of the overall task. At some point, we can indeed tell the robot, "go explore. Report back interesting stuff. We'll be along later."

At the very top level, we need a strategic plan. From the objectives of this plan, a sequence of sub-plans and operations can be derived. We also need contingency cases and the ability to replan, when reality diverges from expectations. The robot needs to be able to, at some level, plan and execute, then evaluate according to success criteria. Task decomposition into relevant sub-tasks is an area that is essential for the more advanced telerobotic systems.

Telerobots are designed to operate in distant or remote areas. They extend the operating envelope of human workers in space and/or time. For example, an underwater telerobot does not need to surface to replenish its air supply, and when it does resurface, it doesn't have to do it slowly to avoid "the bends." A robot on the International Space Station can be kept outside, not needing to "suit-up" before attending to repair scenarios. Telerobots extend the domain where useful or critical work can be done. They reduce human labor requirements, and reduce human exposure to hazardous environments. Space is certainly a hazardous environment for humans.

Telepresense is the extension of a human's senses to a remote location. This includes enhancement of the senses, such as a radiation detector, or night vision. Various aspects of the workstation design for the humans determine the effectiveness of the overall system. For example, should the cameras on the robotic system be slaved to the head motion of the operator? Our tools tend to have our capabilities in mind, so we work best with systems like us, bilaterally symmetric, for example. Having a three--armed robotic device, even though it would be very useful in many tasks, would be difficult for the human operate to get used to. From the task standpoint, a big, strong left arm and two dexterous right arms might be the right choice. Telecontrol refers to the manipulation of the remote system by the human operator, either by direct control of the remote mechanisms, or by higher level directives.

Teleoperator systems have high adaptability and low autonomy, due to the person being an integral part of the control loop. This works well for a certain class of problems. Robots tend to have high autonomy for specific, well-defined tasks, but low adaptability. We can get to autonomous systems by added intelligence and perception to robotic systems, or adding a supervisory mode to the telerobotic systems. NASA has sent robotic systems into Space since the 1970's. It could be argued that any spacecraft is a tele-robot system, but we will take a narrower view.

Teleoperation of a servicer from the Earth's surface is on the edge of feasible, because of the communication delays. Teleoperation from the ISS is also a valid option.

NASA Experience

The NASA Center assigned to Earth orbiting, non-crewed spacecraft is the Goddard Space Flight Center in Greenbelt, Maryland.

Rendezvous, Docking, and Berthing

These operations refer to catching up with the target spacecraft, and capturing it. The target should remain passive, with attitude control disabled when the servicing vehicle is within grappling distance. Then a series of proximity operations (prox ops) begin. If the target is not cooperative, the operations get a whole lot more complicated. Although both spacecraft are traveling at a very high velocity in orbit, the servicer can match velocity with the target.

A problem if the servicer is using gas thrusters is possible contamination of optical surfaces on both spacecraft. The whole task will go a lot easier if the target spacecraft is designed to be serviced. This means it would have a grapple fixture, and it would have an external fuel fitting with a valve and a standard connector. In addition, it may have a modular architecture like GSFC's Multimission Modular Spacecraft, when there were three main modules, attached to a triangular structure. These were for power, attitude control, and command and data handling. Before there was any way to do it, these made the spacecraft serviceable, by allowing the change-out of the modules. When the Shuttle became available, and in-orbit servicing was feasible, this enabled the SMM and HST servicing missions. On HST, the astronauts went above and beyond what the HST was designed to have replaced, showing the versatility of a human on-site.

Docking in orbit was demonstrated by a Gemini mission, and was a key component of the Apollo lunar missions. Both in Earth orbit and in lunar orbit, it was used for the lunar landings. Docking was achieved between an Apollo capsule in Earth orbit, and a Soyuz spacecraft on the Apollo-Soyuz Mission. You can see what this looked like at the Smithsonian's Air & Space Museum in Washington, D. C.

The Russian Igla was a radio-based automated docking system first used on a Soyuz in 1967. It was used on all Russian capsules and space stations.

The later Soyuz-15 mission had to be aborted, because the automated docking failed, and there was no manual backup. The replacement docking systems was called Kurs, It is currently in use on the ISS. It is also used on the European Automated Transfer Vehicle, launched on the Ariane-V. It has three times the capacity of the Russian Progress cargo carrier. Five ATV's were built. Besides their cargo function, the ATV can boost the ISS's orbit.

For Skylab, after the third crewed visit, there were still enough consumables for a fourth visit, and the onboard systems were holding up well. The launch vehicles, the capsules, the astronauts were all available. But, Skylab needed to be re-boosted, and it was big and ungainly, and experienced a lot of orbital decay. A proposed approach was the Teleoperator Retrieval System (TRS) launched with the Shuttle to boost the orbit. Unfortunately, the orbit decayed faster, and the Shuttle was not ready in time. Skylab reentered and burned, and a big chunk hit Australia.

There are now International Standards for Docking, covering crewed and automated spacecraft.

GSFC's Multi-mission Modular Spacecraft

The multi-mission modular spacecraft was conceived at GSFC to solve the cost problem of building each spacecraft in a custom fashion. The MMS was designed as a set of common functionality in a "Bus" configuration, that could handle the majority of requirements of scientific payloads. The science payload would interface to the MMS via a set of standard mechanical and electrical interfaces. The MMS would provide power, communications, attitude, and orbit control The MMS consisted of a framework structure, with 3 modules attached. The attitude control box contained sensors and reaction wheels, with interface for thrusters, all under the control of a NASA Standard Spacecraft computer (NSSC-1). Quite a few missions used the MMS design, including the Solar Maximum Mission (SMM), the International Ultraviolet Explorer (IUE), the Upper Atmosphere Research Satellite (UARS), Landsat -4 and -5, and the Extreme Ultraviolet Explorer (EUVE). MMS missions were compatible with the Delta launch vehicle from the Kennedy Space Center, or Vandenburg Air Force Base in California, for polar orbits. The MMS was also compatible with the Space

Shuttle, with the addition of the Flight Support System (FSS) discussed next.

The FSS

The Flight Support System was a structure designed to hold spacecraft built to the MultiMission Modular Spacecraft design in the Space Shuttle's cargo bay. It could be used to hold a spacecraft for orbital delivery, to return a spacecraft from orbit, and to hold a spacecraft for repair by EVA astronauts. The FSS provided a secure mechanical hold on the spacecraft, and could also hold spare MMS modules.

One proposed mission for the FSS would have used a Shuttle mission launched into polar orbit from Vandenburg Air Force Base to place the Landsat-D (prime) spacecraft in orbit, and retrieve and return with the on-orbit Landsat-D. This mission never occurred, since the Shuttle was never launched into polar orbit.

One very important mission that the FSS did enable was the capture and repair of the Solar Maximum Mission, using the Shuttle and EVA Astronauts.

The SMM Repair Mission

The Solar Maximum Mission satellite was designed to investigate Solar phenomena, particularly solar flares. It was launched on February 14, 1980.

In January 1981, three fuses in the SMM's attitude control system failed, causing it to rely on its magnetic torquers to maintain attitude. In this mode, only three of the seven instruments were usable, as the others required the satellite to be accurately pointed at the Sun. The use of the satellite's magnetic torquers prevented the satellite from being used in a stable position and caused it to "wobble" in its orbit.

Although not unique in this endeavor, the SMM was notable in that its useful lifetime was significantly increased by the direct intervention of a manned space repair mission. During STS-41-C in 1984, the Space Shuttle Challenger intercepted the SMM, maneuvering it into the shuttle's payload bay for maintenance and repairs. SMM had been fitted with a shuttle "grapple fixture" so that the shuttle's robot arm could grab it. During the mission, the SMM's entire attitude control system module and the electronics module for the an instrument were replaced, and a gas cover was placed over another instrument. SMM was the first on-orbit servicing mission in history. The ARM was teleoperated from the Shuttle's aft deck, working alongside EVA astronauts.

The success of the SMM repair demonstrated beyond a doubt the feasibility of servicing a spacecraft in orbit, but at a high level of complexity, involving a Shuttle mission, and trained astronauts. These repairs were successfully completed, adding five years to the satellites working life. The spacecraft reentered the atmosphere and burned in December of 1989, taking some of the author's best flight software with it.

Interestingly, a similar repair exercise was conducted at GSFC's Robotics Lab after the SMM mission, involving a SMM mock-up, and a large industrial PUMA robot arm, operated in telerobotic fashion by GSFC Robotics Branch, code 714.

Reference

Austin, Edmund; Fong, Chung P. "Teleoperated Position Control of a PUMA Robot, 1987 avail: NASA Technical Reports Server, Doc. ID 19890000733.

The Hubble Repairs

There were five servicing missions to the Hubble Space Telescope between 1993-2009. These covered the addition of adaptive optics to correct the main mirror flaw, change-out of some instruments, replacing failed components, and updating the flight computer. This was all made possible by the Shuttle's "Arm," the remote manipulator system (RMS), a tele-robotic device.

The RMS is teleoperated from the Aft flight deck of the Shuttle. It is normally stowed along the sill of the cargo bay, on the shuttle's left side. It is 50 feet long, and can handle a mass of 60,000 pounds. It has been demonstrated to be safe when used in conjunction and proximity to humans (astronauts) also doing servicing tasks. It is designed to be jettison-able if control fails on-orbit, so the cargo bay doors of the Shuttle can be closed for re-entry.

The RMS had a 2-axis shoulder joint where it attaches to the Shuttle, a single axis elbow, and a 3-axis wrist. The end of the RMS is called the end effector, and is designed to mate with a specific fixture on spacecraft or ORU's.

Designed for zero-G operation, the RMS cannot support its own weight on Earth. An air-bearing floor is used for 2-axis simulations. There were full-scale hydraulic analogs at Johnson Space Center and Goddard Space flight Center, but these could only manipulate full size but inflatable payloads. RMS operations are planned with simulation software.

Servicing Mission One in 1993 involved 7 astronauts, the Shuttle's telerobotic arm, and a hundred specialized tools. The arm, operated from the shuttle's aft deck, was used to capture the spacecraft, and maneuver it onto the servicing platform (Flight Support System) in the Shuttle bay. This was possible because the bus side of the HST used the MultiMission Modular Spacecraft (MMS) architecture. Among other things, a co-processor was added to the Rockwell DF-224 flight computer. The co-processor had dual redundant 80386/80387 processor/numeric processor pairs, each with 1MB of RAM and 256kB EEPROM, plus 384kB of non-alterable permanent ROM. The repair mission was a success.

Mission two in 1997, used the same procedures, and replaced some instruments and a tape recorder with a new solid state memory unit.

Mission 3A went in December of 1999, and responded to the failure of 3 of the 6 onboard gyros. The set of six were replaced, as well as a fine guidance sensor, and the main computer. The old computer, a DF-224, was replaced by a new unit, some 20 times faster, and with 6 times the memory. It had three rad-hard Intel 486 processors running at 25MHz, each with 2MB of SRAM and 1MB of EEPROM. It is still operating as of this writing.

Mission 3B in 2002, brought a new instrument, an improved cooler for one of the instruments, and a change-out of the solar panels.

The Shuttle Columbia disaster almost spelled the end of further servicing missions. A study in 2004 by GSFC came to the conclusion that a fully robotic servicing mission was not currently feasible. A new NASA administrator remove the ban on STS servicing missions. In the mean time, the Hubble's main data handling unit failed, bringing science to an abrupt stop. Service Mission 4 replaced the faulty unit in 2009, and added two additional instruments. They also installed the Soft Capture and Rendezvous System, which will enable future robotic missions. Good data from HST is still flowing to the Space Telescope Science Institute on the Johns Hopkins University campus in Baltimore.

NASA-GSFC SSCO, Satellite Servicing Capabilities Office

This group at NASA's Goddard Space Flight Center consists of a large team of veterans of on-orbit servicing missions to the Hubble Space Telescope. These missions had been carried out by EVA Astronauts from the Space Shuttle. Satellite Servicing has been in development at NASA since 1976. NASA sponsored a series of Workshops on the topic in 2010 and 2012. These brought together the NASA team with commercial and academic groups doing similar work from around the world.

This group consists of the largest body of hands-on experience with on-orbit maintenance operations in the world. The heritage of the SMM and Hubble repairs along with the design of serviceable spacecraft and custom tools is part of the group's heritage.

The Shuttle, not capable of reaching Geosync altitude, did retrieve two spacecraft stuck in a lower orbit lower orbit, and returned them to the ground for repair and relaunch. These were the Palapa B2 and Westar-6. Both satellites shared the same platform. There was a failure in the boost motor that would have taken them to geosynchronous orbit. The mission was funded in part by Lloyd's of London, who would otherwise have had to pay for replacements.

RESTORE Project

The RESTORE Project is a public-private partnership to expand the NASA technology of on-orbit servicing to geosynchronous altitudes. It will result in the Restore-L servicer spacecraft.

The stated goals are to:

• Advance the state of robotic servicing technology to enable the routine servicing of satellites that were not designed with servicing in mind

• Position the U.S. to be the global leader in in-space repair, maintenance, and satellite disposal

• Help to enable a future U.S. industry for the servicing of satellites

• Enable the full commercial utilization of NASA-developed technology supporting satellite servicing activities.

This was to be accomplished in the 2018-2023 time frame, an ambitious schedule. NASA sees it's role as developing and demonstration the technology, then letting a commercial entity operate the service. This will initially be a government-industry partnership arrangement, with NASA supplying the initial technology. The mechanism will be a CRADA – a Cooperative Research and Development Agreement between NASA and the selected commercial vendor. These agreements define roles, responsibilities, and ownership. NASA supplies Intellectual Property, technology resources, and expertise to the commercial partner, who will take the ball and score with it. The commercial entity will provide financing for a commercial product and service that will benefit NASA, the Military, the Commercial Satellite sector, and other nations.

NASA's Intellectual Property in this context includes certain patents related to Satellite Services, developed tools, a robotic system with dual, 7-degree of freedom arms, Flight Software, technology for on-orbit fuel transfer, The Space Cube advanced space computer, teleoperation workstations, mission integration and testing of the flight element, launch and operations support, and object recognition software.

Space Systems Loral will take the current technology and bring it to an operational state, affix it to a spacecraft, launch the spacecraft, and conduct the mission.

Upon a successful mission, Loral will have exclusive rights to use the NASA servicing and repair technology in the commercial sector. This is a good deal.

Restore would potentially provide life extension servicing over a range of candidate client satellites. Specific on-orbit servicing capabilities include:

- Remote Survey: visually inspect, record and evaluate client satellite external conditions
- Relocate: re-position client satellite to another orbital location
- Refuel: transfer propellant to/from a client satellite
- Repair: fix degraded, malfunctioning, or inoperative satellite
- Replace: replace degraded, malfunctioning or inoperative satellite components

How does the RSV itself get resupplied and repaired? There are plans for an orbiting re-supply of the RSV vehicle, and it could, in theory, be repaired by another RSV.

Sections of the RPM technology and operations are being tried out on the ISS, using the Dextre robot. Dextre has a toolbox of relevant tools for the servicing operation, including a wire cutter, a multi-function tool, a safety cap tool, and a nozzle tool.

Like Dextre, the RESTORE-L will have two robotic arms. The spacecraft will include an autonomous relative navigation system, designed to fly in close proximity to the target spacecraft

The RESTORE-L Mission will attempt to refuel Landsat-7, currently in low polar orbit, around 2025.

The OMV

The Orbital Maneuvering Vehicle was a planned "space tug" that would be used to capture satellites and bring them to the Space Station for repair and servicing. This never came to pass, partially due to the dangers inherent to the station itself.

Robotic Refueling Mission

The Robotic Refueling Mission (RRM) is a joint NASA/Canadian Space Agency Project to test hardware and techniques for refueling spacecraft in orbit. This will include spacecraft that were not specifically designed to be serviced, or refueled. At this point, not every satellite in orbit has been specifically designed for on-orbit service.

The on-orbit demonstration will be done at the International Space Station. An RRM module weighing 250 kilograms would be mounted outside the station habitable area. It will contain a fluid transfer experiment, using some 1.7 liters of ethanol. Inside the module will be four purpose built tools for testing. These include a wire cutter and (thermal) blanket manipulation tool, a safety cap removal tool, a multi-function tool, and a nozzle tool.

The RPM package was delivered to the Station on Shuttle Mission STS-135, the last mission. It was removed from the shuttle cargo bay by 2 astronauts, and placed on a temporary platform. Later, the Space Stations arm assembly moved it to the Express Logistics Carrier-4.

Personnel at the Johnson Space Center in Houston will operate the Station's Dextre Telerobot, which consists of two dexterous arms. They will use the tools from the RRM module, which latch onto the end of the arms. Additional tools and task boards will be sent to the station later.

In 2012, the RRM had shown that remotely controlled telerobots could perform precise servicing tasks in space, in low-clearance working spaces. A fluid transfer was accomplished in January 2013. All of these operations had been extensively tested and verified on the ground.

The RPM tasks are:

1. Launch Lock Removal and Vision - The Dextre robot releases the "launch locks" on the four RRM servicing tools. These locks kept the tools secure within the RRM module during the shuttle Atlantis' flight to the International Space Station. Then Dextre's cameras image the hardware in both sunlight and darkness, providing data to develop machine vision algorithms that work against harsh on-orbit lighting.

2. Gas Fittings Removal - Marking the first use of RRM tools on orbit, Dextre uses the tools to remove the fittings that many spacecraft have for the filling of special coolant gases.

3. Refueling - After snipping lock wires and removing caps, Dextre is able to access a fuel valve similar to those commonly used on satellites today and transfer liquid ethanol through a sophisticated robotic fueling hose, completing a first-of-its-kind robotic refueling event.

4. SMA (Sub-miniature type A) Cap Removal - Dextre removes the coaxial radio frequency (RF) connector caps that terminate and protect the RF connector while the satellite is in orbit. Access to these connectors would

allow a robotic servicer to plug into the data systems of a satellite and better diagnose an internal issue.

5. Screw Removal - Dextre will robotically unscrew satellite bolts (fasteners). RRM draws from its experience with the Hubble Space Telescope servicing mission in its use of a small cage to guide the tool tip and ensure that no fasteners float away.

6. Thermal Blanket Manipulation - Dextre slices off thermal blanket tape and folds back a thermal blanket to access the contents underneath.

Reference: http://ssco.gsfc.nasa.gov/rrm_tasks.html

NASA's DART

NASA's DART (Demonstration of Autonomous Rendezvous Technology) Program in 2005 involved an autonomous rendezvous between the DART spacecraft, and MublCOM (MUltiple Paths Beyond Line-of-site COMmunication). Unfortunately, that test resulted in a collision.

The Military Experience

There is little known about the military's classified satellite programs. DARPA's Phoenix program is a satellite project to recycle retired satellite parts into new on-orbit assets. This is quite a bit more complex than refueling. They are looking to harvest solar arrays, antennae, or other accessible stuff. They say, "...Phoenix is truly all about going up to retired, non-cooperative, non-controlled satellites that have been left for dead in geosync graveyard orbits." In 2016, the project was renamed "Robotic Servicing of Geosynchronous Satellites." A new spacecraft was formed to address these needs, Space Infrastructure Servicing, by MacDonald, Dettwiler and Associates of Canada.

The U. S. Naval Academy built a low-cost, 3U cubesat to demonstrate in-orbit repairs. It is equipped with two 7 DOF robotic arms, as well as advanced imaging.

DARPA Orbital Express

This DARPA project to develop "a safe and cost-effective approach to autonomously service satellites in orbit" was conducted in conjunction with NASA's Marshall Space Flight Center. The servicing satellite was called ASTRO, and the prototype serviceable satellite was termed NEXTsat. The mission was launched in 2007. Ball Aerospace built NEXTSat, and MacDonald-Dettwiler contributed the Orbital Express Demonstration Manipulator System (OEDMS). ASTRO was built by Boeing. NASA-Marshall Space Flight Center provided the Advanced Video Guidance System. VACCO Industries was responsible for the refueling mechanism, for hydrazine. Sierra Nevada Corp. furnished the docking mechanism. Besides a refueling, the mission would change out a power-ORU. Multiple transfers were completed over the following months in orbit. The OEDMS had a 6-DOF robot arm with vision. Both spacecraft were deactivated after the successful mission, and reentered.

The Robotic Servicing of Geosynchronous Satellites (RSGS) Project uses a commercial spacecraft bus, with DARPA contributing the "toolkit" including the robotic arms. It would be capable of high-resolution inspection, repairs, upgrades, and re-location/reboost. Space Systems Loral was selected in 2017 as the commercial partner. SSL pulled out of the project in 2019, due to costs.

NextSat

NEXTsat, or Next Generation Satellite and Commodities Spacecraft, was the target for on-orbit servicing, and was designed specifically to be serviced in space. That involves many factors including accessibility. NEXTSat was built by Ball Aerospace, and was launched in Low Earth Orbit in 2007. It weighed around 500 pounds, and was designed for a four month mission duration. After it was separated from Astro, it was decommissioned

Astro

The ASTRO servicing satellite was launched with NEXTSat, along with 3 other satellites. It was designed to demonstrate servicing and refueling of NEXTSat. Built by Boeing, Astro weighed more than a ton. After successful operations, it was separated from the NEXTSat, and deactivated. It has subsequently re-entered the atmosphere.

The Astro system included a robotic arm with six degrees of freedom, and a vision system for autonomous operation.

The Commercial Spacecraft Experience

Commercial Spacecraft, mostly in the communications satellite business, greatly benefit from repairs and refueling satellites in orbit. Refueling is life-extension technology. Another area being addressed is launch or deployment anomalies, where the satellite is sent to the wrong orbit by a partially faulty launch vehicle, and deployment anomaly's, where the solar arrays or antennas are not correctly deployed. Part of the servicing scenario would include imaging, so that the cause of a problem could be analyzed.

In addition, servicing satellites could move a zombie-sat from the operation orbit to the graveyard orbit, assuring it does not collide with or interfere with operational units.

Intelsat

MacDonald Dettwiler (MDA) of British Columbia, Canada announced in 2011, a $280 million agreement with Intelsat for the servicing of on-orbit satellites via a space-based service vehicle to be provided by MDA.

The Intelsat organization, operating since 1964, had a fleet of 52 communications satellites, which the world depends on for voice and data traffic. Intelsat has over 600 Earth stations in 150 countries. This infrastructure is expensive to build, launch, and maintain. Getting the most usage out of the infrastructure is critical.

In early 2011, two commercial spacecraft providers announced plans to provide new autonomous/teleoperated unmanned resupply spacecraft for servicing other unmanned spacecraft. Both of these servicing spacecraft will be docking with satellites that were designed neither for docking, nor for in-space servicing.

Influenced by the 2007 Orbital Express mission, a U.S. Government effort to test in-space satellite servicing with two vehicles designed for on-orbit refueling and subsystem replacement, two companies announced new commercial satellite servicing missions that will require docking of two unmanned vehicles.

Space Infrastructure Servicing (SIS) is a spacecraft being developed by Canadian aerospace firm MacDonald, Dettwiler and Associates (MDA), maker of Canadarm on the ISS, to operate as a small-scale in-space refueling depot for communication satellites in geosynchronous orbit.

As of March 2011, Intelsat has agreed to purchase one-half of the 2,000 kilogram propellant payload that an MDA Corporation spacecraft satellite-servicing demonstration project would take to geostationary orbit. Catching up in orbit with four or five Intelsat communication satellites, a fuel load of 200 kilograms of fuel delivered to each satellite would add somewhere between two and four years of additional service life. A near-end-of-life Intelsat satellite will be moved to a graveyard orbit 200 to 300 kilometers above the geostationary belt where the refueling will be done, "without consequence" to the Intelsat business.

As of March 2010, the business model was still evolving. MDA "could ask customers to pay per kilogram of fuel successfully added to the satellite, with the per-kilogram price being a function of the additional revenue the operator can expect to generate from the spacecraft's extended operational life."

The plan is that the fuel-depot vehicle would maneuver to several satellites, dock at the target satellite's apogee-kick motor, remove a small part of the target spacecraft's thermal protection blanket, connect to a fuel-pressure line and deliver the propellant. "MDA officials estimate the docking maneuver would take the communications satellite out of service for about 20 minutes."

The challenge is, to capture satellites with no grapple fixture, and to refuel satellites not designed to be refueled. The newer satellites in the series need to be reworked to be serviceable.

Orbital-ATK

Orbital-ATK and U. S. Space, in a joint venture called ViviSat, are addressing the servicing problem with the *Mission Extension Vehicle* for commercial spacecraft. As of this writing, the company has received orders for two vehicles from Intelsat. The MEV provides rendezvous, proximity operations, and docking capabilities. The vehicle is targeted to fuel replacement, inspection, repair, replacement of parts and assemblies, or it can provide auxiliary propulsion, navigation, or power to the target spacecraft. Of course, the target has to be serviceable, meaning it was designed to be serviced in the first place.

You don't buy a car where the gas filler door is welded shut. HST and SMM were mostly designed for servicing, although the astronauts on-site used specially designed tools, and ingenuity.

MEV is designed for a 15-year lifetime on orbit. It is based on the Orbital GEOStar Core, and incorporates both gas jet and electric propulsion. The concept dates to 2011. The first launch was in 2020. MEV-1 docked successfully with Intelsat-901 on February 25. The Intelsat bird was 19 years old and working fin, but ran out of stabilization fuel. The MEV had launched in October, 2019, and was tested extensively in orbit. It went up to Intelsat-901 in its graveyard orbit, some 300 km above geostationary. Docking was Feb. 25, 2020. It is expected that 901 will go back into service in March. The Intelsat was not designed with servicing in mind. A second MEV is in construction. It uses the same satellite bus as the communications satellites. Intelsa-901 and -907 are next in line.

Orbital ATK is now Northrop Grumman Innovation Systems.

International efforts

The early European efforts included the Orbital Recovery Corporation, and Orbital Satellite Services, Ltd. They would attach to a satellite and provide propulsion, navigation, and guidance services. This would allow for adjusting the orbit, due to decay or launch vehicle error, and tow zombie-sats out of the way of operating spacecraft.

European OLEV (Orbital Life Extension Vehicles) included Orbital Recovery Group's ConeXpress. This project did not proceed to an orbital test, but a new venture called Orbital Satellite Services, LTD (OSSL) was formed by some of the key players. Their vehicle was based on an ESA design from the Small Missions for Advanced Research and Technology (SMART) Program. This was scheduled for an in-orbit demo in 2011, but never happened.

ESA

The ConeXpress was a concept for geosynchronous satellite life extension. It would launch on an Ariane-5, to sync orbit. It was primarily designed to refuel. The mission was not implemented.

German Efforts

The German Space Agency, DLR, developed the designs for the DEOS (Deutsche Orbitale Servicing Mission) robotic spacecraft. It was supposed to be capable of capturing an "un-cooperative" target, which would then put into a "destructive re-entry". This is for tidying up orbital debris. No money was forthcoming for this project.

Canadian Space Robotics

Canada has been the major supplier of space robotic/telerobotic hardware, including the arms on the shuttles and the ISS. This work was originally done at Spar Aerospace of Edmonton, Ontario. This company was a vendor to the Canadian Space Agency (*Agence spatiale canadienne*), headquartered in Saint-Hubert, Quebec. It was formed in 1990, and does joint projects with NASA in the United States, and ESA, in Europe. Seventeen Canadian astronauts have flown in space missions. Spar is now a part of MacDonald Detweiler, operating as MD Robotics.

Other Approaches

Swarms

A different approach to repair robotics uses collections of smaller co-operating multi-robotic systems that can combine their efforts and work as ad-hoc teams on problems of interest.

This is based on the collective or parallel behavior of homogeneous systems. This covers collective behavior, modeled on biological systems. Examples in nature include migrating birds, schooling fish, and herding sheep. A collective behavior emerges form interactions between members of the swarm, and the environment.

In Swarm robots, the key issues are communication between units, and cooperative behavior. The capability of individual units odes not much matter; it is the strength in numbers. Ants and other social insects such as termites, wasps, and bees, are models for robot swarm behavior. Self-organizing behavior emerges from decentralized systems that interact with members of the group, and the environment. Swarm intelligence is an emerging field, and swarm robotics is in its infancy.

Wrap-up

It is cheaper to repair than to replace, in most cases. Systems with multiple problems might required replacement, but that just leaves us with a disposal problem. Satellites are expensive assets, and getting them to orbit is more expensive. Unlike most assets, they are not serviced. You don't buy a new car, when your old one has a burned out taillight, do you?

By developing a Orbital Servicing Infrastructure, we can extend the life of very expensive resources, and clean up some of the orbital clutter that is endangering current and future space assets.

Bibliography

AIAA, *Design for On-Orbit Spacecraft Servicing*, 1991, AIAA Guide G-042, 1991, ISBN-1563470284.

Albu-Schaffer, Alin "DLR's Robotic Technologies for Space Debris Mitigation and On-Orbit Servicing, Institute of Robotics and Mechatronics, avail: http://www.unoosa.org/pdf/pres/stsc2013/2013iaf-05E.pdf

Autonomous Mission for On-Orbit Servicing, Final Report, 2014, International Space University, avail: https://www.academia.edu/8731611/AMOOS_Autonomous_Mission_for_On-Orbit_Servicing_._Co-authoring_the_report_to_the_Space_Studies_Program_2014_of_the_International_Space_University_Montreal_CA_2014

Benedict, Bryan, "Rationale for Need of In-Orbit Servicing Capabilities for GEO Spacecraft," AIAA SPACE 2013 Conference and Exposition, AIAA SPACE Forum, (AIAA 2013-5444) avail: https://arc.aiaa.org/doi/abs/10.2514/6.2013-5444.

Biesbroek, Robin, *Active Debris Removal in Space: How to Clean the Earth's Environment from Space Debris,* 2015, ISBN-1508529183.

Billig, Frederick S. *Design for On Orbit Spacecraft Servicing* (Conference Proceeding Series), American Institute of Aeronautics and Astronautics, 1992, ISBN-10: 1563470063.

Cepollina, F. J. "Economic and Technical Aspects of Repair, Servicing, and Retrieval of Low Earth Orbit Free Flying Spacecraft," Oct, 1982, AIAA Paper 82-1810, avail: https://sspd.gsfc.nasa.gov/images/NASA_Satellite_Servicing_Project_Report_2010A.pdf

Cepollina, Frank J.; Burns, Richard D.; Holz, Jill M.; Corbo, James E.; Jedhrich, Nicholas M. "Method and associated apparatus for capturing, servicing, and de-orbiting earth satellites using robotics," 2009, US-Patents - 7,513,460; 7,513,459; 7,438,264; 7,293,743; 7,240,879. avail: Google Patents.

Cepollina, Frank J., *The History of Satellite Servicing at GSFC*, GSFC, Systems Engineering Seminar, May 2, 2006, PowerPoint, https://ses.gsfc.nasa.gov/ses_data_2006/060502_Cepollina_Abstract.htm.

Cepollina, Frank J., Remote Servicing of Free-flying Spacecraft, 1975, avail: https://ntrs.nasa.gov/archive/nasa/casi.ntrs.nasa.gov/19770022806.pdf

Cruzen, Carig (ed), Schmidjuber, Michael (ed), Lee, Yiung H. (ed), Kim, Bangyeop (ed) *Space Operations: Contributions from the Global Community*, (March, 2017), Springer, ASIN-B06XY3G21T.

Currie, Nancy J.; Peacock, Brian INTERNATIONAL SPACE STATION ROBOTIC SYSTEMS OPERATIONS – A HUMAN FACTORS PERSPECTIVE NASA, Johnson Space Center, National Space Biomedical Research Institute, Houston, Texas avail: https://www.nasa.gov/centers/johnson/pdf/486042main_I SSRoboticsHumanFactorsPerspective.pdf

Davinic, Nick; Arkus, Alan; Chappie, Scot; Greenberg, Joel "Cost-benefit Analysis of on-orbit satellite servicing," J. Reducing Space Mission Cost, 1998, Vi issue 1, avail: https://link.springer.com/article/10.1023/A:100995590948.

DARPA, Parish, Joseph, Robotic Servicing of Geosynchronous Satellites (RSGS). Avail: https://www.darpa.mil/program/robotic-servicing-of-geosynchronous-satellites

http://www.esa.int/Our_Activities/Human_Spaceflight/International_Space_Station/European_Robotic_Arm.

Ferster, Warren (2013-05-17). "DARPA Cancels Formation-flying Satellite Demo". *Space News*. Avail:https://spacenews.com/35375darpa-cancels-formation-flying-satellite-demo/

Foust, Jeff "Robots and Hubble: a bad idea?", Nov. 1, 2004, The Space Review, article 257/1, avail: http://www.thespacereview.com/article/257/1

Gefke, Gardell G. *Advances in Robotic Servicing Technology Development,* 2015, ASIN-B01BB1ETJE.

Gordon, Scott A. *Design Guidelines for Robotically Serviceable Hardware,* NASA/GSFC, 1988, NASA Tech Memo 100700, avail: https://spacenews.com/35375darpa-cancels-formation-flying-satellite-demo/

Greenberg, Joel S. "Economic Principles Applied to Space Industry Decisions," Vol 201, January 1, 2003, AIAA, ISBN-156347607X.

Griffin, Thomas J.; Setwart, William N.; *"The GSFC Flight Support System for on-orbit satellite servicing,"* Jan 1988, AIAA PAPER 88-0448, avail: https://arc.aiaa.org/doi/abs/10.2514/6.1988-448

Holcomb, Lee "The NASA Automation and Robotics Technology Program," 1985, In-Space Research,

Technology and Engineering Workshop, available, ResearchGate.net

Horsham, Gary "Examination of Prospects for Satellite Servicing, A Common Government/Industry Strategy for the Development of Space," NASA/TM-2-1—216937, AIAA 2010-889, 2002. avail: NASA Technical Reports Server, NTRS.nasa.gov

Horsham, Gary A. P.; Gilland, James H. "Establishing a Robotic, LEO-to-Geo Satellite Servicing Infrastructure as an Economic Foundation for Exploration," Nov. 2010, NASA/TM-2010-216937, AIAA-2010-8897, avail: NASA Technical Reports Server, NTRS.nasa.gov

Horsham, Gary A. P. *Envisioning a 21st Century, National, Spacecraft Servicing and Protection Infrastructure and Demand Potential: A Logical Development of the Earth Orbit Economy,* NASA TM—2003-212462, avail: NASA Technical Reports Server, NTRS.nasa.gov.

Horsham, Gary NASA, "Examination of Prospects for Satellite Servicing - A Common Government/Industry Strategy for the Development of Space," NASA Office of Space Flight Advanced Systems, 2002, avail: NASA Technical Reports Server, NTRS.nasa.gov

Joppin, Carole; Hastings, Daniel E. "On-orbit Upgrade and Repair; the Hubble Space Telescope Example," J. Spacecraft and Rockets, Vol. 43, No. 3, May-June 2006 avail: https://arc.aiaa.org/doi/abs/10.2514/1.15496?journalCode=jsr

Korf, Richard E. *Space Robotics*, Carnegie-Mellon University, Robotics Institute, Aug. 1982,CMU-RI-TR-82-10. Avail: https://ri.cmu.edu/pub_files/pub3/korf_richard_e_1982_1/korf_richard_e_1982_1.pdf

Kosmas, C. "On-Orbit-Servicing by HERMES On-Orbit-Servicing System," Policy Robust Planning, AIAA, avail: www.georing.biz/OTHER/GEORING56926.pdf

Kunstadter, Christopher T. W. "Space Insurance: Experience and Outlook A Statistical Review of Volatility", March 2002, United States Aviation Underwriters. Avail: http://slideplayer.com/slide/8288638/

Launius, Roger D.; McCurdy, Howard E. *Robots in Space,* 2008, JHU Press, ISBN 0-8018-8708-9.

Launius, Roger D.; McCurdy, Howard E. *Robots in Space: Technology, Evolution, and Interplanetary Travel,* JHU Press, ASIN-B01FELC04I.

Leete, Stephen J. "Design for On-Orbit Spacecraft Servicing, NASA-Goddard Space Flight Center, avail: https://pdfs.semanticscholar.org/9890/48ef58f9d9860f20579fcce26433aaca6d0c.pdf

Liang, Bin, Li, Cheng, Xue, Lijun, Qiang, Wenyi; "A Chinese Small Intelligent Space Robotic System for On-Orbit Servicing," 2006, Intelligent Robots and Systems, IEEE ISBN 1-4244-0258-1.avail: Researchgate.net

Long, Andrew M.; Richards, Matthew G.; Hastings, Daniel E. "On-Orbit Servicing: A New Value Proposition for Satellite Design and Operation," JOURNAL OF SPACECRAFT AND ROCKETS, Vol. 44, No. 4, July–August 2007, avail: https://arc.aiaa.org/doi/10.2514/1.27117

Moe, Rud V. "Hierarchy of on-orbit servicing interfaces," 1989, ESA, Second European In-Orbit Operations Technology Symposium; p 77-81, avail: https://ntrs.nasa.gov/search.jsp?R=19900014978.

Montgomery, Ray C. (ed); Kaufman, Howard (ed) "Selected topics in Robotics for Space Exploration," 1993, NASA-CP-10131, avail: Google books.

NASA/Goddard Space Flight Center, *On-Orbit Satellite Servicing Study*, Project Report, Oct. 2010. avail:

https://sspd.gsfc.nasa.gov/images/NASA_Satellite%20Servicing_Project_Report_0511.pdf

NASA Facts, Robotic Refueling Mission, FS-2013-03-017-GSFC. Avail: https://sspd.gsfc.nasa.gov/images/RRM_Factsheet.pdf

NASA, Satellite Services Workshop, Volume 1 Final Report, NASA Lyndon B. Johnson Space Center, Houston, Texas, U.S. Department of Commerce, National Technical Information Service, June 1982. avail:https://ntrs.nasa.gov/archive/nasa/casi.ntrs.nasa.gov/19830002905.pdf

NASA, Satellite Services Workshop, Volume 2 Final Report, NASA Lyndon B. Johnson Space Center, Houston, Texas, U.S. Department of Commerce National Technical Information Service, June 1982. avail:

https://sspd.gsfc.nasa.gov/documents/2_Satellite_Services_Workshop_Vol_2_Jun_82.pdf

NASA, Satellite Services Workshop II, Volume 1, NASA Lyndon B. Johnson Space Center, Houston, Texas, U.S. Department of Commerce National Technical Information Service, November 1985. avail:

https://sspd.gsfc.nasa.gov/documents/2_Satellite_Services_Workshop_Vol_2_Jun_82.pdf

NASA, Satellite Services Workshop II, Volume 2, NASA Lyndon B. Johnson Space Center, Houston, Texas, U.S. Department of Commerce National Technical Information Service, November 1985.

NASA, *EVA Tools and Equipment Reference Book*, NASA Lyndon B. Johnson Space Center, Houston, Texas, U.S. Department of Commerce National Technical Information Service, November 1983. avail: https://ntrs.nasa.gov/archive/nasa/casi.ntrs.nasa.gov/19940017339.pdf

NASA, "On-Orbit Servicing Experience - A Compilation of Lessons Learned," NASA On-Orbit Servicing Steering Committee, Advanced Program Development Division, June 1990, avail:https://sspd.gsfc.nasa.gov/library.html

NASA, On-Orbit Satellite Servicing Study, NASA-GSFC, Oct. 2010, avail:https://sspd.gsfc.nasa.gov/images/NASA_Satellite%20Servicing_Project_Report_0511.pd

NASA, Hubble Space Telescope Servicing Mission 4, The Soft Capture and Rendezvous System," NASAFacts, FS-2007-08-088-GSFC (SM4 #05) (rev. 11/07). avail: https://www.nasa.gov/pdf/206046main_SCRS_FS-2007-08-088_12_4.pdf

NASA, Space Station Program, *Robotic Systems Integration Standards*, Vol. 1, Robot Accommodation Requirements, International Space Station Program, Rev. c, Sept. 1998, SSP 30550 V1 R C. avail:

https://sspd.gsfc.nasa.gov/documents/10_Space_Station_Program_Robotic_Systems_Integration_Standards_1998.pdf

NASA, SSSWG Meeting #22, Presentation Materials, Satellite Services System, NASA DOD Industry. November 1989. avail: https://sspd.gsfc.nasa.gov/documents/12_SSWG_Meeting__22_Nov_1989.pdf

NASA, *On-orbit Satellite Servicing Study*, Oct. 2010, avail: https://sspd.gsfc.nasa.gov/images/nasa_satellite%20servicing_project_report_0511.pdf

Nof, Shimon Y. *Springer Handbook of Automation*, 2009, ISBN-3540788328.

Ollendorf, Stanford "Robotic servicing on Earth orbiting satellites," 1993. NASA. Johnson Space Center, Sixth Annual Workshop on Space Operations Applications and Research (SOAR), Avail: ntrs.nasa.gov

Ollendorf, Stanford "Robotic Deployment and Servicing of Scientific Payloads on the Lunar Surface," 1991,

Research & Technology Report, N94-14791, avail: NTRS.NASA.gov

Ollendorf, Stanford *"Robotics program at the NASA Goddard Space Flight Center,"* 1989, ESA, Second European In-Orbit Operations Technology Symposium; p 13-21, avail: NTRS.NASA.gov

Ousley, Gilbert W.; "Returned Solar Max Hardware Degradation Study Results NASA-GSFC, N-89-23537, avail: NTRS.NASA.gov

Patterson, Linda P. "On-Orbit Maintenance Operations Strategy for the International Space Station - Concept and Implementation," https://ntrs.nasa.gov/search.jsp?R=201000425252019-05 05T20:00:22+00:00Z

Pelton, Joseph *New Solutions for the Space Debris Problem*, Springer, 2015, ISBN-331917150X.

Reed, Benjamin *Overview of NASA's In Space Robotic Servicing,* 2015, ASIN-B01BB1EDJK.

Reynerson, C. M "Spacecraft modular architecture design for on-orbit servicing,", Ball Aerospace. & Technology Corp., Boulder, CO, Aerospace Conference Proceedings, 2000, (Volume 4) American Institute of

Aeronautics & Astronautics, February 1992, ISBN-1563470284.

Roberts, Brian John *Challenges of In Space Robotic Servicing,* 2015, ASIN-B01BB1EDZ4.

Schenker, P. S. "NASA Research and Development for Space Telerobotics," IEEE Transitions on Aerospace and Electronic Systems, Sept 1988, v 24 n 5 pp 523-534.

Shayler, David J., Harland, David M. *The Hubble Space Telescope: From Concept to Success*, 2016, ISBN-1493928260.

Shayler, David J., Harland, David M. *Enhancing Hubble's Vision: Service Missions That Expanded Our View of the Universe*, 2016, ISBN-3319226436

Stakem, Patrick H. "Advanced Computational Architecture for Flight Telerobotic Servicers", Satellite Services Workshop IV, June 21- 23, 1988, Johnson Space Center, Texas. Avail: ntrs.nasa.gov.

Stakem, Patrick H. *Robots and Telerobots in Space Applications*, 2011, PRRB Publishing, ASIN B00571MJRM.

Stolarik, Ellen G. (ed); Littlefield, Ronald G (ed); Beyer, David S. (ed) *Proceedings of the 1987 Goddard conference on Space Applications of Artificial Intelligence (AI)* and Robotics, May 13-14, 1987, NASA/GSFC, avail:ntrs.nasa.gov

Triolo, Jack, *Fender Bender, System Failure Case Studies*, Vol 2, No. 7, Sept 2008. NASA Safety Center, System Failure Case Studies, http://pbma.nasa.gov/

Truszkowski, Walt; Rouff, Christopher; Karlin, Jay; Rash, James; Hallock, Harold; Hinchley, Michael; *Autonomous and Autonomic Systems: With Applications to NASA Intelligent Spacecraft Operations and Exploration Systems*, Springer; December 8, 2009, ISBN-1846282322.

United States Aviation Underwriters, *Space Insurance: Experience and Outlook*, March 2002, avail: https://slideplayer.com/slide/8288638/

Waltz, Donald M. *On-Orbit Servicing of Space Systems,* 1993, ISBN-089464002X.

Whipple, Art *A Comparison of Human and Robotic Servicing of the Hubble Space Telescope*, NASA-GSFC, Oct 6, 2009.

Whipple, Fred "Meteorites and Space Travel", 1947 Astronomical Journal, 52, 131.

Wolf, Thomas "Deutsche Orbitale Servicing Mission," Space-Administration of the German Aerospace Center, robotics.estec.esa.int/ASTRA/Astra2011/Presentations/Plenary%202/04_wolf.pdf

Wolf, Thomas, "The In-flight Technology Demonstration of German's Robotics Approach to Dispose Malfunctioned Satellites," Space Administration of the German Aerospace Center (DLR). Avail:

http://robotics.estec.esa.int/ASTRA/Astra2011/Papers/00/FCXNL-11A06-2145230-1-2145230wolf.pdf

Xu, Yangsheng; Kanade, Takeo *Space Robotics: Dynamics and Control,* Springer, 1993, ISBN 1461365953

Resources

NASA Orbital Debris Program Office; http://orbitaldebris.jsc.nasa.gov

In-Orbit Servicing archives, https://spacenews.com/tag/in-orbit-servicing/

http://ssco.gsfc.nasa.gov/robotic_refueling_mission.html

NASA, *Renewing Solar Science, the Solar Maximum Repair Mission*, N86-24714, avail: NTRS.nasa.gov.

NASA Tech Reports Library, http://ntrs.nasa.gov (2600 references on OnOrbit Servicing)

http://www.spacenews.com/article/satellite-builders-not-enthusiastic-about-orbit-servicing-project#.UfKM8m2wW5w

http://acuriousguy.blogspot.com/2011/04/backgrounder-for-on-orbit-satellite.html

Satellite Services Workshop Vol 1, Final Report, N83-11175, Johnson Space Center, Jun,1982 NASA-TM-84873.

Satellite Services Workshop Vol 2, Final Report, N83-11175, Johnson Space Center, Jun, 1982 NASA-TM-84874.

Satellite Services Workshop II , Vol 1, N83-11175, Johnson Space Center, Nov 1985, JSC-20677.

Satellite Services Workshop II , Vol 2, N83-11175, Johnson Space Center, Nov 1985, JSC-20677 vol 2.

EVA Tools and Equipment Reference Book, NASA-TM-109350, JSC-20466, Nov. 1993.

http://www.spacenews.com/article/satellite-builders-not-enthusiastic-about-orbit-servicing-project#.UfKM8m2wW5w

http://acuriousguy.blogspot.com/2011/04/backgrounder-for-on-orbit-satellite.html

"Intelsat Picks MacDonald, Dettwiler and Associates Ltd. for Satellite Servicing

http://www.fas.org/spp/military/docops/usaf/2020/app-h.html

http://www.extremetech.com/extreme/146495-darpa-shows-ff-satellite-scavenging-project-phoenix-tech

"The GEO Graveyard May Not Be Permanent," Staff, Tech Space, Nov. 2010, www.spacedaily.com.

https://sspd.gsfc.nasa.gov/robotic_refueling_mission.html

http://robonaut.jsc.nasa.gov/default.aspA Survey of Multi-Robot Cooperation in Space,

avail: https://pdfs.semanticscholar.org/1.pdf

NASA GSFC Library, gsfcir.gsfc.nasa.gov

DARPA, Roesler, Dr. Gordon, Robotic Servicing of

Geosynchronous Satellites, avail: https://www.darpa.mil/program/robotic-servicing-of-geosynchronous-satellites

http://spectrum.ieee.org/automaton/robotics/space-robots/.WJzFIylBe5c.linkedin

The Near Term Future of On-Orbit Servicing is Robotic, avail:

http://spaceref.biz/commercial-space/the-near-term-future-of-on-orbit-servicing-is-robotic.html

http://gcd.larc.nasa.gov/projects/robotic-satellite-servicing

zombie- sat   http://www.milsatmagazine.com/story.php?number=152118614 Galaxy-15

spacenews.com/on-orbit-satellite-servicing-the-next-big-thing-in-space

https://sspd.gsfc.nasa.gov/images/nasa_satellite%20servicing_project_report_0511.pdf

MDA, Robotics and On-orbit Servicing, avail:

https://mdacorporation.com/isg/robotics-automation/space-based-robotics-solutions/robotics-solutions/robotics-and-on-orbit-servicing

International Docking System Standard (IDSS) Interface Definition Document, avail:

https://ntrs.nasa.gov/search.jsp?R=20170001546

AIAA, "On-Orbit Servicing Will Lower Costs and Increase Satellite Life Spans," avail:

https://space.aiaa.org/On-OrbitServicing/

Intelsat General corp. "Rationale for Need of In-Orbit Servicing Capabilities for GEO Spacecraft,"

avail:https://www.intelsatgeneral.com/wp-content/uploads/2015/04/GEO-In-orbit-Servicing-Challenges6_2.pdf

International Docking System Standard (IDDS) Interface Definition Document (IDD), Rev. E, 2013

NASA, Roscosmos, ESA, JAXA, avail: http://www.internationaldockingstandard.com/download/IDSS_IDD_Revision_E_TAGGED.pdf

On-Orbit Servicing, Executive Summary, On-Orbit Servicing Corporation, Nov. 3003

wikipedia, various

Glossary

Actuator – device which converts a control signal to a mechanical action.
AMODS – Autonomous Mobile On-orbit Diagnostic System
APAS - Androgynous Peripheral Attach System; Androgynous Peripheral Assembly System.
APDS - Androgynous Peripheral Docking System
AR&D – Autonomous Rendezvous and Docking.

ASAT – anti-satellite (weapon).
ASIN – Amazon Standard Inventory Number
ASTRO – Autonomous Space Transport Robotic Operations – U.S. Tech demo satellite.
ATP – authority to proceed.
ATV – Automated Transfer Vehicle, European.
BAA – Broad Agency Announcement (U. S. Government)
CPU – central processing unit
CERS – crew emergency rescue system
CRADA – Cooperative Research and Development Agreement (U.   S. Government and industry)
Cryogenic – very low temperature.
Cryote – Cryogenic Orbital Testbed.
CSA – Canadian Space Agency.
DARPA – (U. S.) Defense Advanced Research Projects Agency.
DART – Demonstration (of) Autonomous Rendezvous Technology.
Dextre - Dexterous Manipulator robot arm, Canadian, on Space Station.
DFOSS – Design for on-orbit spacecraft servicing.
DLR – German Space Agency (Deutsches Zentrum für Luft- und   Raumfahrt)
DoD – (U. S.) Department of Defense.
DOF – degrees of freedom.
ELV – Expendable Launch Vehicle.
ESA – European Space Agency
EVA – Extra Vehicular Activity- involving an Astronaut with suit and maneuvering unit in space.
FAR – (US) Federal Acquisition Regulations

FISO – future in-space operations
FPP – Firm Fixed Price (Contract)
FREND – Front-end Robotics Enabling Near-term Demonstration (DARPA).
FSS – Flight Support System, structure in Space Shuttle bay to hold spacecraft.
FTS – Flight Telerobotic Servicer.
GEO – geosynchronous Earth orbit, 22,236 miles.
GHz – giga (109) hertz.
Giga - 109 or 230.
GOES – NASA/NOAA Geostationary Operational Environmental Satellite
GPU – graphics processing unit.
GNFIR - GSFC Natural Feature Image Recognition System
Graveyard orbit – a place to park end-of-life satellites.
Gray - unit of radiation, =100 rad
GSFC – Goddard Space Flight Center, Greenbelt, Maryland. NASA Center for unmanned spacecraft near Earth.
IDD – Interface Definition Document.
IDSS – International Docking System Standard
Igla – Soviet docking system for Soyuz.
Intelsat = International Telecommunications Satellite Organization.
IP – Intellectual Property
ISBN – International Standard Book Number.
ISS – International Space Station.
JAXA - Japan Aerospace Exploration Agency
LEO – Low Earth Orbit
LIDS – Low impact docking system.

LSP – NASA launch services program.
LTG – LEO to GEO.
LV – launch vehicle.
MCU – media control unit (touchscreen) on Tesla.
MES – mission extension services,
MEV-1 (Orbital-ATK) Mission Extension Vehicle-1
MMS – (NASA/GSFC) MultiMission Modular Spacecraft.
MMU – manned maneuvering unit – for EVA astronauts.
MOOSE – Manned on-orbit servicing equipment.
MSFC – Marshall Space flight Center, Huntsville, Alabama.
Mublcom - MUltiple Paths Beyond Line-of-site COMmunication.
NASA – National Aeronautics and Space Administration (USA).
NDS – NASA Docking System
NEXTsat – Next Generation Satellite and Commodities Satellite.
NIST – National Institutes of Standards and Technology.
NOAA – National Oceanographic and Atmospheric Administration. (USA)
NRL – U.S. Naval Research Center.
NSSC-1 NASA Standard Spacecraft Computer-1.
OEDMS - Orbital Express Demonstration Manipulator System.
OLEV - Orbital Life Extension Vehicles
Open source – methodology for hardware or software development with free distribution and access.
OSSL - Orbital Satellite Services, LTD.
ORU – Orbital Replacement Unit.

ProxOps – proximity operations.
RCS – robot control system; reaction control system
PDGF – Power Data Grapple Fixture, on the Space Station.
POD – (DARPA) payload orbital delivery
POES – Polar orbiting environmental satellite.
RFI – Request for Information; radio frequency interference.
RNS – Relative Navigation System
RRM – Robotic Refueling Mission.
RSAT-P Repair Satellite Prototype
RSGS – Robotic servicing of geosynchronous satellites.
RSV – RESTORE Servicing Vehicle; robotic servicing vehicle
RWS – Robotic Work Station, on Space Station.
SAFR – simplified Aid for crew rescue
SARAH – Self Adaptive Robotic Auxiliary Hand, (on Dextre)
SCM – Soft Capture Mechanism.
SIS – Space Infrastructure Servicing
SMM – Solar Maximum Mission, an MMS mission.
SPDM – Special Purpose Dexterous Manipulator on Space Station,   aka Dextre
SSCO – Satellite Servicing Capabilities Office, NASA, GSFC.
SSL – Space Systems Loral.
STS – Space Transportation System (USA) Shuttle.
TDRS – Tracking and Data Relay Satellite.
Telecheric – control of robot at a distance. Teleoperated, with   Telepresense
Telerobot – a robotic system with a human in the loop.

ULA – United Launch Alliance, commercial launch services company.
Ullage – the fuel left in an "empty" tank.
Zombie-Sat – dead satellite posing a danger to other spacecraft

If you enjoyed this book, you might also be interested in some of these.

Stakem, Patrick H. *16-bit Microprocessors, History and Architecture*, 2013 PRRB Publishing, ISBN-1520210922.

Stakem, Patrick H. *4- and 8-bit Microprocessors, Architecture and History*, 2013, PRRB Publishing, ISBN-152021572X,

Stakem, Patrick H. *Apollo's Computers*, 2014, PRRB Publishing, ISBN-1520215800.

Stakem, Patrick H. *The Architecture and Applications of the ARM Microprocessors*, 2013, PRRB Publishing, ISBN-1520215843.

Stakem, Patrick H. *Earth Rovers: for Exploration and Environmental Monitoring*, 2014, PRRB Publishing, ISBN-152021586X.

Stakem, Patrick H. *Embedded Computer Systems, Volume 1, Introduction and Architecture*, 2013, PRRB Publishing, ISBN-1520215959.

Stakem, Patrick H. *The History of Spacecraft Computers from the V-2 to the Space Station*, 2013, PRRB Publishing, ISBN-1520216181.

Stakem, Patrick H. *Floating Point Computation*, 2013, PRRB Publishing, ISBN-152021619X.

Stakem, Patrick H. *Architecture of Massively Parallel Microprocessor Systems*, 2011, PRRB Publishing, ISBN-1520250061.

Stakem, Patrick H. *Multicore Computer Architecture,* 2014, PRRB Publishing, ISBN-1520241372.

Stakem, Patrick H. *Personal Robots*, 2014, PRRB Publishing, ISBN-1520216254.

Stakem, Patrick H. *RISC Microprocessors, History and Overview,* 2013, PRRB Publishing, ISBN-1520216289.

Stakem, Patrick H. *Robots and Telerobots in Space Application*s, 2011, PRRB Publishing, ISBN-1520210361.

Stakem, Patrick H. *The Saturn Rocket and the Pegasus Missions, 1965,* 2013, PRRB Publishing, ISBN-1520209916.

Stakem, Patrick H. *Visiting the NASA Centers, and Locations of Historic Rockets & Spacecraft,* 2017, PRRB Publishing, ISBN-1549651205.

Stakem, Patrick H. *Microprocessors in Space*, 2011, PRRB Publishing, ISBN-1520216343.

Stakem, Patrick H. Computer *Virtualization and the Cloud*, 2013, PRRB Publishing, ISBN-152021636X.

Stakem, Patrick H. *What's the Worst That Could Happen? Bad Assumptions, Ignorance, Failures and Screw-ups in Engineering Projects, 2014,* PRRB Publishing, ISBN-1520207166.

Stakem, Patrick H. *Computer Architecture & Programming of the Intel x86 Family, 2013,* PRRB Publishing, ISBN-1520263724.

Stakem, Patrick H. *The Hardware and Software Architecture of the Transputer*, 2011,PRRB Publishing, ISBN-152020681X.

Stakem, Patrick H. *Mainframes, Computing on Big Iron*, 2015, PRRB Publishing, ISBN- 1520216459.

Stakem, Patrick H. *Spacecraft Control Centers*, 2015, PRRB Publishing, ISBN-1520200617.

Stakem, Patrick H. *Embedded in Space,* 2015, PRRB Publishing, ISBN-1520215916.

Stakem, Patrick H. *A Practitioner's Guide to RISC Microprocessor Architecture*, Wiley-Interscience, 1996, ISBN-0471130184.

Stakem, Patrick H. *Cubesat Engineering*, PRRB Publishing, 2017, ISBN-1520754019.

Stakem, Patrick H. *Cubesat Operations*, PRRB Publishing, 2017, ISBN-152076717X.

Stakem, Patrick H. *Interplanetary Cubesats*, PRRB Publishing, 2017, ISBN-1520766173 .

Stakem, Patrick H. Cubesat Constellations, Clusters, and Swarms, Stakem, PRRB Publishing, 2017, ISBN-1520767544.

Stakem, Patrick H. *Graphics Processing Units, an overview*, 2017, PRRB Publishing, ISBN-1520879695.

Stakem, Patrick H. *Intel Embedded and the Arduino-101, 2017,* PRRB Publishing, ISBN-1520879296.

Stakem, Patrick H. *Orbital Debris, the problem and the mitigation*, 2018, PRRB Publishing, ISBN-*1980466483.*

Stakem, Patrick H. *Manufacturing in Space*, 2018, PRRB Publishing, ISBN-1977076041.

Stakem, Patrick H. *NASA's Ships and Planes*, 2018, PRRB Publishing, ISBN-1977076823.

Stakem, Patrick H. *Space Tourism*, 2018, PRRB Publishing, ISBN-1977073506.

Stakem, Patrick H. *STEM – Data Storage and Communications*, 2018, PRRB Publishing, ISBN-1977073115.

Stakem, Patrick H. *In-Space Robotic Repair and Servicing*, 2018, PRRB Publishing, ISBN-1980478236.

Stakem, Patrick H. *Introducing Weather in the pre-K to 12 Curricula, A Resource Guide for Educators*, 2017, PRRB Publishing, ISBN-1980638241.

Stakem, Patrick H. *Introducing Astronomy in the pre-K to 12 Curricula, A Resource Guide for Educators*, 2017, PRRB Publishing, ISBN-198104065X.

Also available in a Brazilian Portuguese edition, ISBN-1983106127.

Stakem, Patrick H. *Deep Space Gateways, the Moon and Beyond*, 2017, PRRB Publishing, ISBN-1973465701.

Stakem, Patrick H. *Exploration of the Gas Giants, Space Missions to Jupiter, Saturn, Uranus, and Neptune*, PRRB Publishing, 2018, ISBN-9781717814500.

Stakem, Patrick H. *Crewed Spacecraft*, 2017, PRRB Publishing, ISBN-1549992406.

Stakem, Patrick H. *Rocketplanes to Space*, 2017, PRRB Publishing, ISBN-1549992589.

Stakem, Patrick H. *Crewed Space Stations,* 2017, PRRB Publishing, ISBN-1549992228.

Stakem, Patrick H. *Enviro-bots for STEM: Using Robotics in the pre-K to 12 Curricula, A Resource Guide for Educators,* 2017, PRRB Publishing, ISBN-1549656619.

Stakem, Patrick H. *STEM-Sat, Using Cubesats in the pre-K to 12 Curricula, A Resource Guide for Educators*, 2017, ISBN-1549656376.

Stakem, Patrick H. *Lunar Orbital Platform-Gateway*, 2018, PRRB Publishing, ISBN-1980498628.

Stakem, Patrick H. *Embedded GPU's*, 2018, PRRB Publishing, ISBN- 1980476497.

Stakem, Patrick H. *Mobile Cloud Robotics*, 2018, PRRB Publishing, ISBN- 1980488088.

Stakem, Patrick H. *Extreme Environment Embedded Systems,* 2017, PRRB Publishing, ISBN-1520215967.

Stakem, Patrick H. *What's the Worst, Volume-2*, 2018, ISBN-1981005579.

Stakem, Patrick H., *Spaceports*, 2018, ISBN-1981022287.

Stakem, Patrick H., *Space Launch Vehicles*, 2018, ISBN-1983071773.

Stakem, Patrick H. *Mars*, 2018, ISBN-1983116902.

Stakem, Patrick H. *X-86, 40th Anniversary ed*, 2018, ISBN-1983189405.

Stakem, Patrick H. *Lunar Orbital Platform-Gateway*, 2018, PRRB Publishing, ISBN-1980498628.

Stakem, Patrick H. *Space Weather*, 2018, ISBN-1723904023.

Stakem, Patrick H. *STEM-Engineering Process*, 2017, ISBN-1983196517.

Stakem, Patrick H. *Space Telescopes,* 2018, PRRB Publishing, ISBN-1728728568.

Stakem, Patrick H. *Exoplanets*, 2018, PRRB Publishing, ISBN-9781731385055.

Stakem, Patrick H. *Planetary Defense*, 2018, PRRB Publishing, ISBN-9781731001207.

Patrick H. Stakem *Exploration of the Asteroid Belt*, 2018, PRRB Publishing, ISBN-1731049846.

Patrick H. Stakem *Terraforming*, 2018, PRRB Publishing, ISBN-1790308100.

Patrick H. Stakem, *Martian Railroad,* 2019, PRRB Publishing, ISBN-1794488243.

Patrick H. Stakem, *Exoplanets,* 2019, PRRB Publishing, ISBN-1731385056.

Patrick H. Stakem, *Exploiting the Moon,* 2019, PRRB Publishing, ISBN-1091057850.

Patrick H. Stakem, *RISC-V, an Open Source Solution for Space Flight Computers,* 2019, PRRB Publishing, ISBN-1796434388.

Patrick H. Stakem, *Arm in Space*, 2019, PRRB Publishing, ISBN-9781099789137.

Patrick H. Stakem, *Extraterrestrial Life*, 2019, PRRB Publishing, ISBN-978-1072072188.

Patrick H. Stakem, *Space Command*, 2019, PRRB Publishing, ISBN-978-1693005398.

Cuberovers, A Synergy of Technologys, 2020, PRRB Publishing, ISBN-979-8651773138.

Robotic Exploration of the Icy moons of the Gas Giants. 2020, PRRB Publishing, ISBN- 979-8621431006

History & Future of Cubesats, PRRB Publishing, ISBN-979-8649179386.

Hacking Cubesas, Cybersecurity in Space, 2020, PRRB Publishing, ISBN-979-8623458964.

Powerships, Powerbarges, Floating Wind Farms: electricity when and where you need it, 2021, PRRB Publishing, ISBN-979-8716199477.

Hospital Ships, Trains, and Aircraft, 2020, PRRB Publishing, ISBN-979-8642944349.

CubeRovers, a Synergy of Technologys, 2020, ISBN-979-8651773138

Exploration of Lunar & Martian Lava Tubes by Cube-X, ISBN-979-8621435325.

Robotic Exploration of the Icy moons of the Gas Giants, ISBN- 979-8621431006.

History & Future of Cubesats, ISBN-978-1986536356.

Robotic Exploration of the Icy Moons of the Ice Giants, by Swarms of Cubesats, ISBN-979-8621431006.

Swarm Robotics, ISBN-979-8534505948.

Introduction to Electric Power Systems, ISBN-979-8519208727.

Centros de Control: Operaciones en Satélites del Estándar CubeSat (Spanish Edition), 2021, ISBN-979-8510113068.

Exploration of Venus, 2022, ISBN-979-8484416110.

Patrick H. Stakem, *The Search for Extraterrestial Life,* 2019, PRRB Publishing, ISBN-1072072181.

The Artemis Missions, Return to the Moon, and on to Mars, 2021, ISBN-979-8490532361.

James Webb Space Telescope. A New Era in Astronomy, 2021, ISBN-979-8773857969.

Riverine Ironclads, Submarines, and Aircraft Carriers of the American Civil War, 2019, ISBN- 978-1089379287.

www.ingramcontent.com/pod-product-compliance
Lightning Source LLC
Chambersburg PA
CBHW030444220526
45464CB00006B/2410